AF270383

ACCIDENTAL
SCIENCE DISCOVERIES

VELCRO

Kenny Abdo

Fly!
An Imprint of Abdo Zoom
abdobooks.com

abdobooks.com

Published by Abdo Zoom, a division of ABDO, P.O. Box 398166, Minneapolis, Minnesota 55439. Copyright © 2024 by Abdo Consulting Group, Inc. International copyrights reserved in all countries. No part of this book may be reproduced in any form without written permission from the publisher. Fly!™ is a trademark and logo of Abdo Zoom.

Printed in the United States of America, North Mankato, Minnesota.
102023
012024

Photo Credits: Abdo, Alamy, AP Images, Getty Images, Shutterstock
Production Contributors: Kenny Abdo, Jennie Forsberg, Grace Hansen
Design Contributors: Candice Keimig, Neil Klinepier, Colleen McLaren

Library of Congress Control Number: 2023938016

Publisher's Cataloging-in-Publication Data

Names: Abdo, Kenny, author.
Title: Velcro / by Kenny Abdo
Description: Minneapolis, Minnesota : Abdo Zoom, 2024 | Series: Accidental science discoveries | Includes online resources and index.
Identifiers: ISBN 9781098284145 (lib. bdg.) | ISBN 9781098284862 (eBook) | ISBN 9781098285227 (Read-to-Me eBook)
Subjects: LCSH: Fasteners--Juvenile literature. | Industrial design--Juvenile literature. | Serendipity in science--Juvenile literature. | Inventions--Juvenile literature. | Discoveries in science--Juvenile literature.
Classification: DDC 500--dc23

TABLE OF CONTENTS

VELCRO

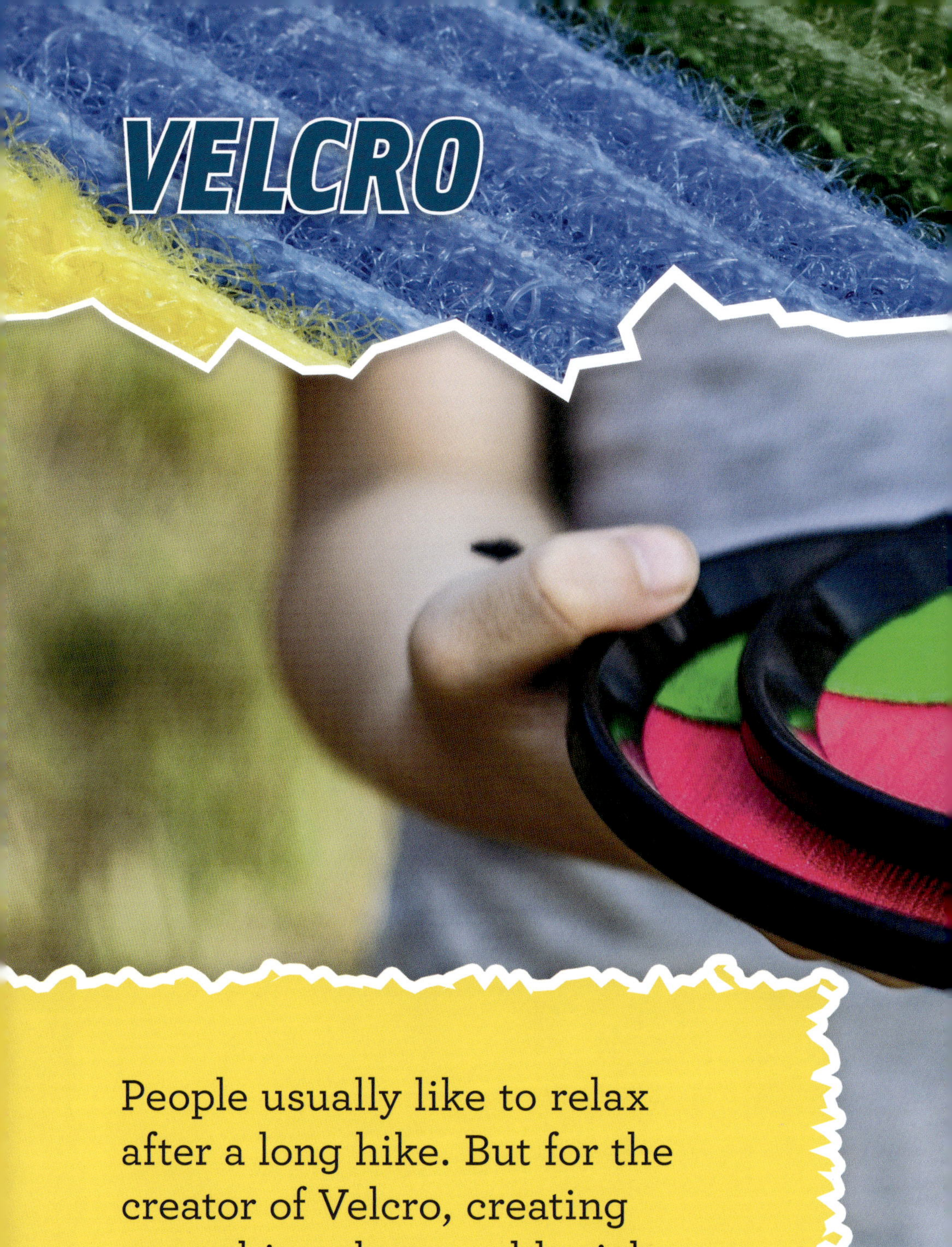

People usually like to relax after a long hike. But for the creator of Velcro, creating something that would stick with the world sounded better!

THE ACCIDENT

Swiss **engineer** George de Mestral was on a walk with his dog in 1941. When they returned home, De Mestral had noticed **burrs** clinging to his clothes. Interested, he studied them closer.

De Mestral discovered that the **burrs** had tiny hooks that caught onto loops in the fabric of his clothing. Inspired by this natural method, he set out to create a similar fastening system.

THE DISCOVERY

After years of tests, De Mestral developed a **prototype**. Combining the words "velvet" and "crochet," Velcro was born. De Mestral started the company Velcro S.A. in 1952.

At first, not many companies believed in Velcro. That all changed in 1961, when **NASA** took a chance on the new product. Used for space suits and keeping instruments secure, Velcro's reputation skyrocketed!

3 4 5 6 7 8
35 36 37 38 39 40 41 42
Mercury

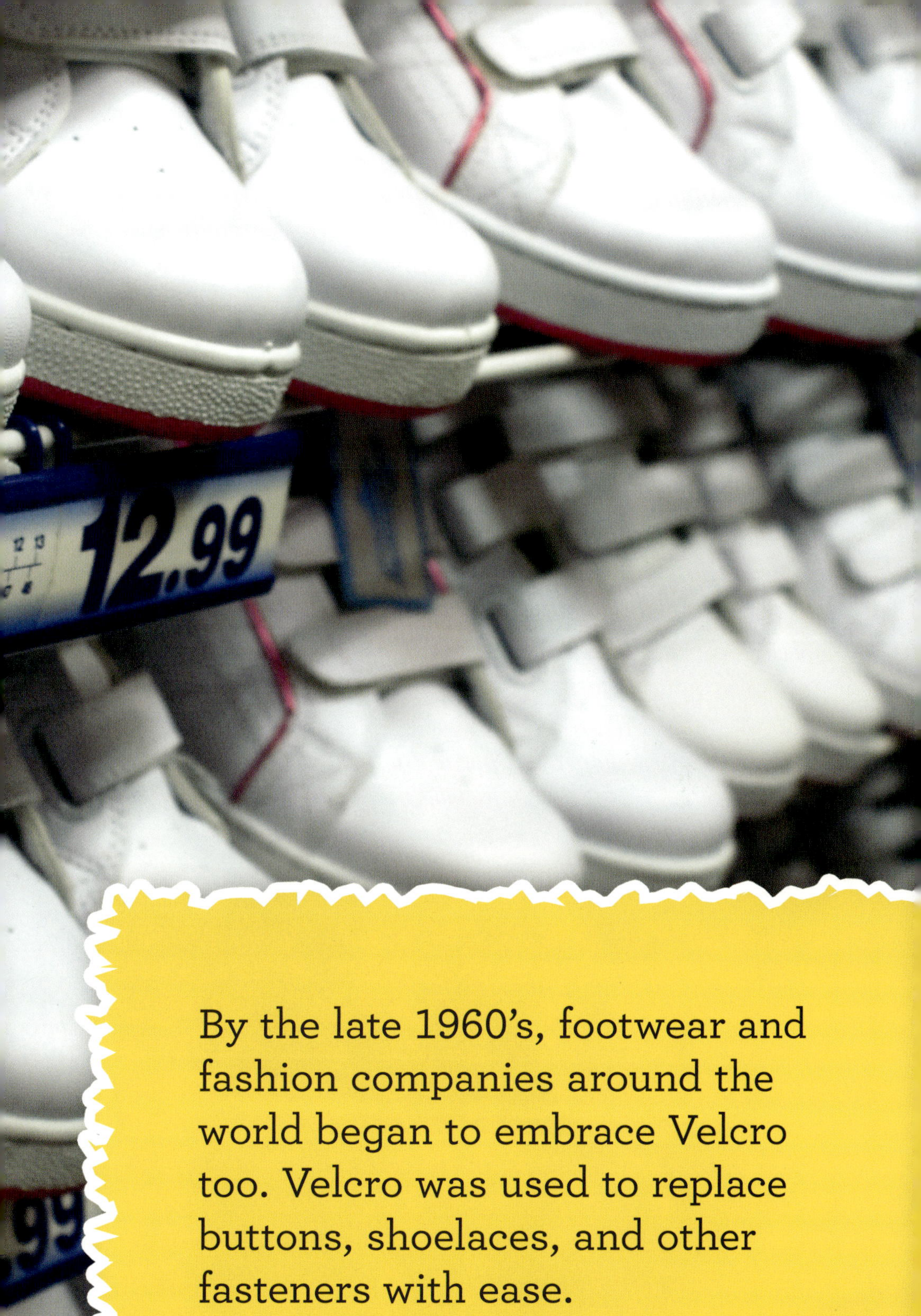

By the late 1960's, footwear and fashion companies around the world began to embrace Velcro too. Velcro was used to replace buttons, shoelaces, and other fasteners with ease.

120
140
160
100
180
80
200
60
220
40
240
20
260
0
280
300
mmHg
Arte

Over the years, Velcro continued to be refined. Companies developed many different types for certain uses. Velcro was used in athletics, diapers, and packaging.

VELCRO
BRAND
SEW ON TAPE
RUBAN À COUDRE
SEW 'EM!
À COUDRE
I'M MADE FROM 70% RECYCLED MATERIALS
DURABLE & WASHABLE
FABRIQUÉES À 70% EN MATÉRIAU RECYCLÉ
RÉSISTANT ET LAVABLE
2½ IN. x ¾ IN.
(6.3 CM x 1.9
8 STRIPS
BANDE

As environmental concerns grew, **sustainable** solutions became more important. In 2020, Velcro Companies began making products with more recyclable and **biodegradable** materials.

From its accidental discovery in the **Swiss Alps** to its widespread use around the world, Velcro has certainly secured its place in history!

GLOSSARY

biodegradable – the ability for a material to be broken down naturally by the organisms in an ecosystem.

burr – a prickly seed case or flower head that clings to clothing and animal fur.

engineer – a person who creates or builds structures and devices using science and math.

NASA – short for National Aeronautics and Space Administration. NASA is run by the US government to study Earth, our solar system, and outer space.

prototype – an early model of an idea or product.

sustainability – the practice of protecting the natural environment while driving innovation and not compromising the way of life.

Swiss Alps – the highest and most densely populated mountain range in Europe.

ONLINE RESOURCES

Booklinks
NONFICTION NETWORK
FREE! ONLINE NONFICTION RESOURCES

To learn more about Velcro, please visit **abdobooklinks.com** or scan this QR code. These links are routinely monitored and updated to provide the most current information available.

INDEX